SCIENCE FACTORY

ELECTRICITY & BATTERIES

MICHAEL
FLAHERTY

PowerKiDS
press.

New York

*T*HE WORKSHOP

BEFORE YOU START any of the projects, it is important that you learn a few simple rules about the care of your science factory.

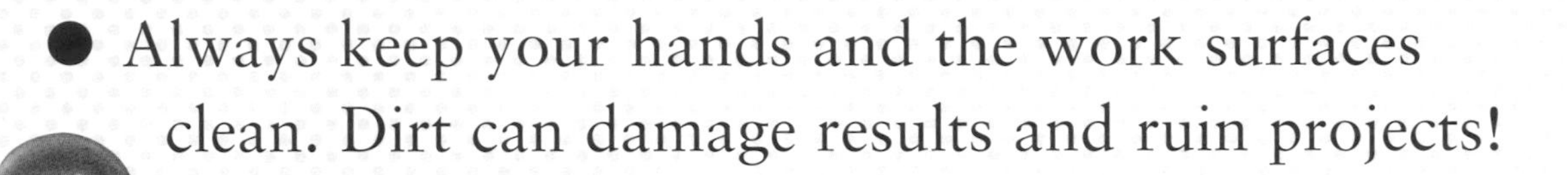

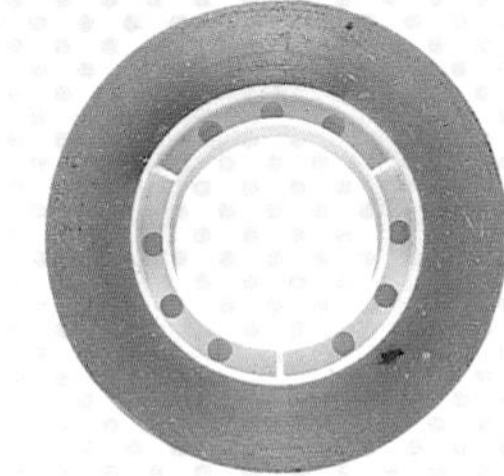

- Always keep your hands and the work surfaces clean. Dirt can damage results and ruin projects!

- Read the instructions carefully before you start each project.

- Make sure you have all the equipment you need for the project (see checklist opposite).

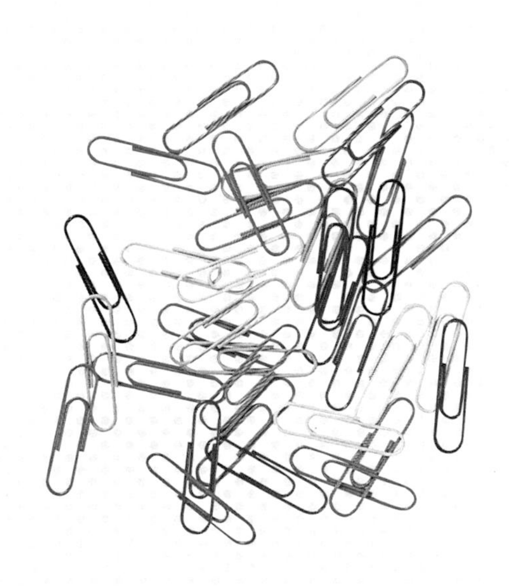

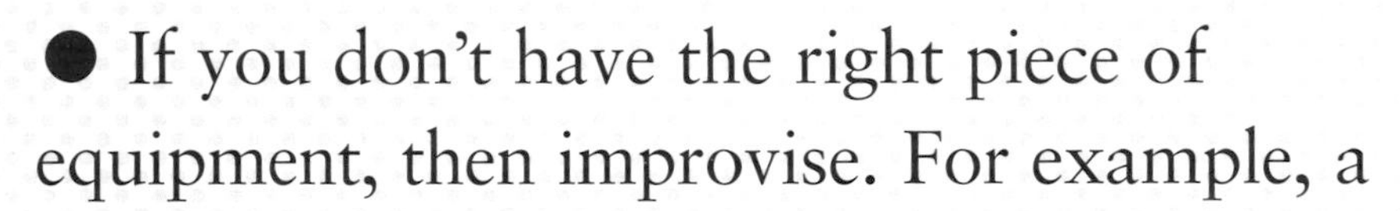

- If you don't have the right piece of equipment, then improvise. For example, a dishwashing-liquid bottle will work just as well as a plastic drink bottle.

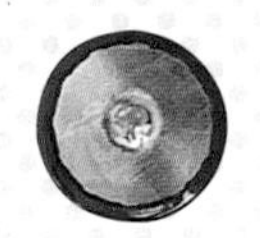

- Don't be afraid to make mistakes. Just start again — patience is very important!

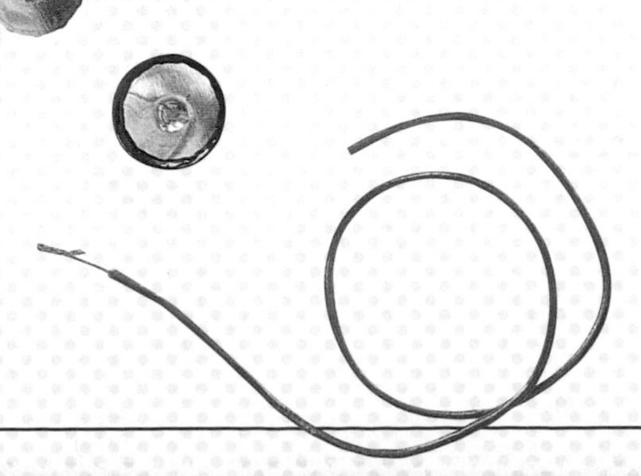

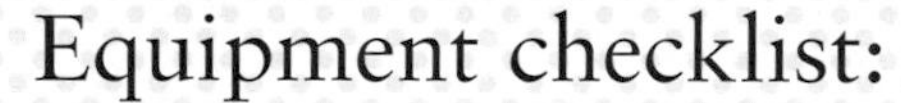

Equipment checklist:

- Scissors and glue
- Table-tennis balls
- Wool cloth
- String and sticks
- Foil candy wrappers and aluminum foil
- 6 volt lightbulbs and sockets
- Paper, cardboard, and tissue paper
- Modeling clay and adhesive vinyl
- Insulated wire
- Plastic drinking straws
- Paper clips and safety pins
- 1.5 volt batteries
- Thumbtacks and nails
- Styrofoam blocks and boards
- Metal coathanger
- Plastic lids and pencil lead
- Glass jar with cork lid
- Copper coins and brass paper fasteners

WARNING:
Some of the experiments in this book need the help of an adult. Always ask a grown-up to give you a hand when you are using sharp objects, such as thumbtacks, or electrical appliances!

STATIC ELECTRICITY

THERE ARE TWO MAIN FORMS OF ELECTRICITY — static (still) and current (flowing). Some materials do not let electricity pass through them, but a static electrical charge is produced on their surface when they rub against certain other materials. When you take off your sweater, you may hear a crackling sound as you produce static electricity. Make frogs jump by static electricity.

WHAT YOU NEED
Tissue paper
Colored cardboard
Table-tennis ball
String
Stick
Wool cloth
Scissors

FROLICKING FROGS

1 Fold a piece of tissue paper a number of times and cut out the shape of a frog. This way you can cut out more than one frog at the same time.

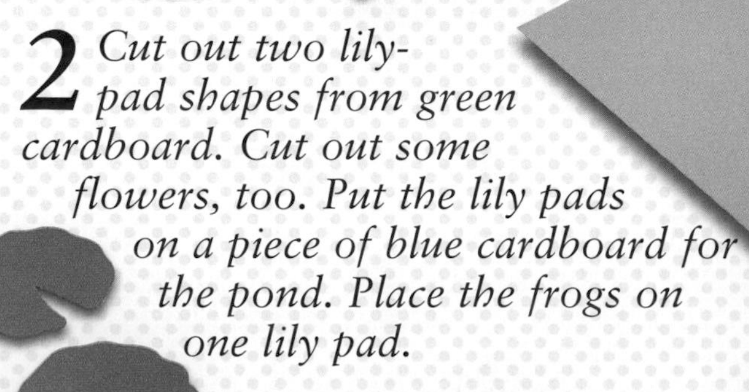

2 Cut out two lily-pad shapes from green cardboard. Cut out some flowers, too. Put the lily pads on a piece of blue cardboard for the pond. Place the frogs on one lily pad.

3 Cut a bird shape out of yellow cardboard. Thread string through the bird and a table-tennis ball to join them together.

4 Tie the other end of the string to the end of the stick. Make sure the bird rests on top of the ball.

5 Rub the table-tennis ball against something made of wool. This gives it an electrical charge.

WHY IT WORKS

The atoms that make up materials have negatively charged electrons and positively charged protons. When you rub the ball, you rub off electrons, leaving the ball with a positive charge. Because unlike charges attract each other, the positively charged ball attracts the paper frogs which have a negative charge in relation to the ball.

STICKY BALLOONS

Rub a balloon against something wool. Hold it against a door or wall and let go. The balloon seems stuck to the door. It is held to the wall by static electricity. The charge slowly disappears. How long does the effect last?

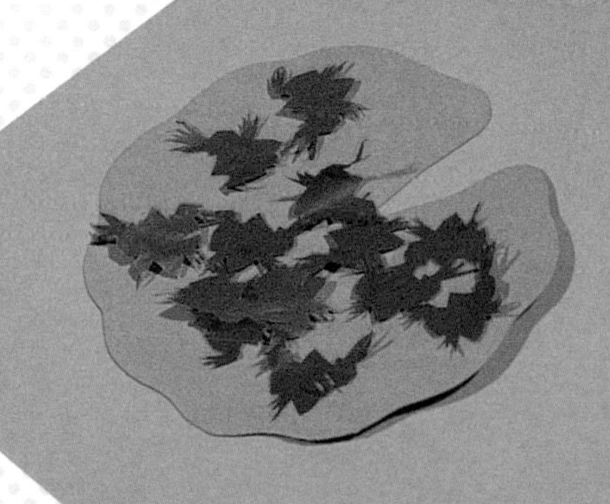

6 Swing the bird and ball over the lily pond. Watch the frogs jump toward the bird. See if the frogs can leap to the other pad.

DETECTING A CHARGE

IN THE LAST PROJECT, we saw how static electricity can be made through rubbing. In 1895, an apparatus called a gold leaf electroscope was invented to detect static electrical charge. You can build an apparatus like a primitive electroscope that shows you if a material is charged with static electricity or not.

WHAT YOU NEED

Bare wire
Aluminum foil
Jar with cork lid
Thin foil candy wrapper
Balloon

PRIMITIVE ELECTROSCOPE

1 Push a piece of wire through a large cork. Bend one end of the wire to make an "L" shape.

2 Roll a piece of aluminum foil into a ball and push it onto the top of the wire that's sticking out of the cork.

3 Fold a piece of thin foil from a candy wrapper in half and rest it on the bottom of the L-shaped wire. Put the cork in the jar, sealing the wrapper inside the jar.

PUSHY BALLOONS

Blow up two balloons and rub them on a wool cloth to charge them. Then use thread to hang them up close to each other. Because you have rubbed electrons from the balloons onto the wool, the balloons are both positively charged. As a result, they repel one another and push apart.

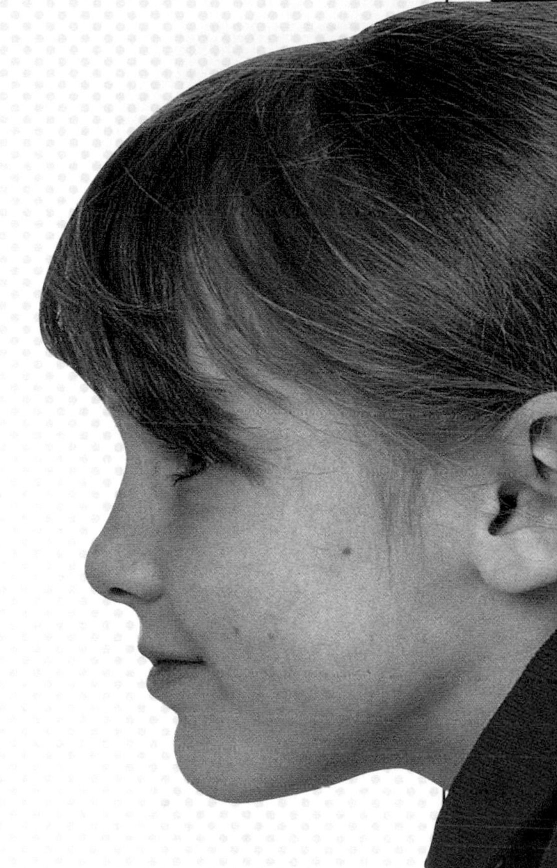

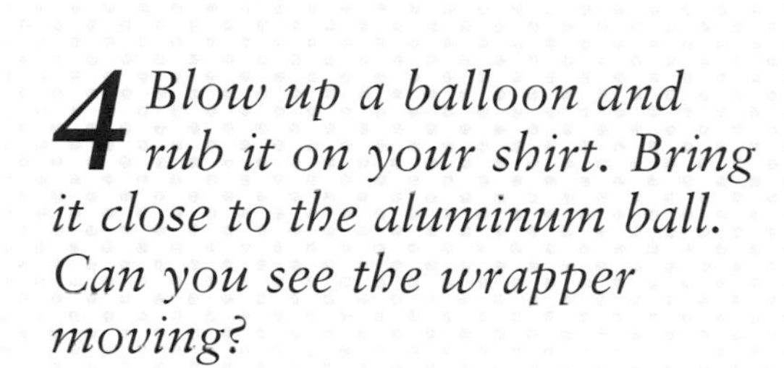

4 *Blow up a balloon and rub it on your shirt. Bring it close to the aluminum ball. Can you see the wrapper moving?*

WHY IT WORKS

Rubbing the balloon rubs off some of its electrons and makes it charged with static. When it comes close to the aluminum foil ball, electrons in the wrapper are attracted to the balloon and move toward it and into the aluminum ball. This leaves the metal foil in the jar positively charged. The two wings of foil try to repel each other and push apart, so the foil moves.

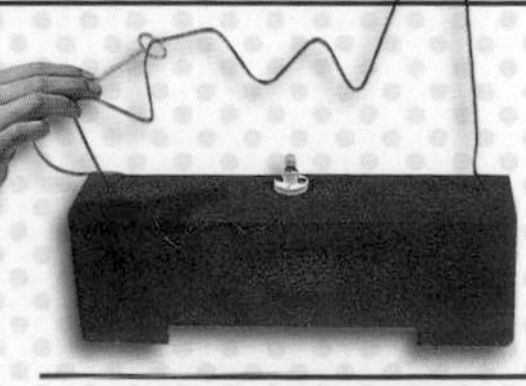

A SIMPLE CIRCUIT

TODAY, WE CAN GENERATE large amounts of electricity. Metal wires and cables carry electricity from power plants to our homes. The electricity travels along these wires like water in a pipe. By switching on a light, you are completing one of these pathways, called a circuit. Electricity now flows through the light. The project below is fun to do, and it lets you set up your own circuit.

WHAT YOU NEED

6-V lightbulb in socket
Insulated wire
Thin bare wire
Drinking straw
Two 1.5V batteries
Styrofoam block
Wire coat-hanger

A STEADY HAND

1 Screw a small 6-V lightbulb into a socket and attach a length of insulated wire to each side of the socket.

2 Make a piece of thin bare wire into a loop. Connect the loop and a long piece of insulated wire together. Thread a plastic drinking straw onto the wire and cover the joint with the straw to form a handle.

3 Ask an adult to open up a wire coat-hanger and bend it into humps and curves. Attach one wire from the lightbulb to one end of the coathanger. Attach the other wire from the lightbulb to two batteries that have been taped together.

4 Attach the wire loop to the other end of the batteries. You can use modeling clay or tape to hold the wires on the ends of the batteries.

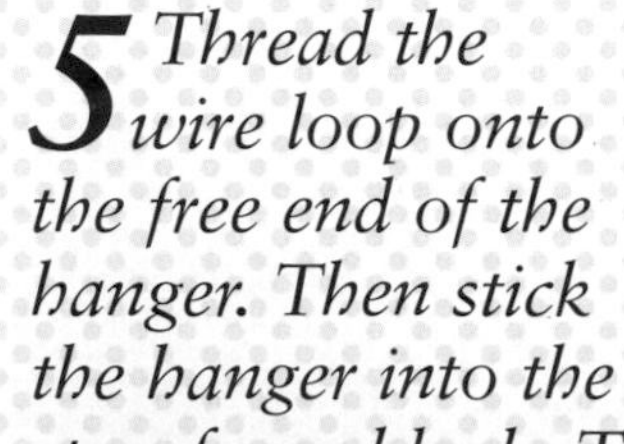

5 Thread the wire loop onto the free end of the hanger. Then stick the hanger into the styrofoam block. Try to move the loop along the coat hanger without their touching.

SHORT CIRCUIT

Set up a simple circuit by connecting bare wires from the lightbulb to the battery. Then lay a metal object, such as a spoon, across the wires. The lightbulb goes out. You have made a short circuit.

If you touch the coat-hanger, the bulb will light up. See how far you can get!

WHY IT WORKS

The electrons won't flow around the circuit until it is complete. By touching the loop to the hanger wire, you complete the circuit, and the bulb lights up as the electrons flow. Electrons flow from the negative end of the battery around the wire to the positive end of the battery.

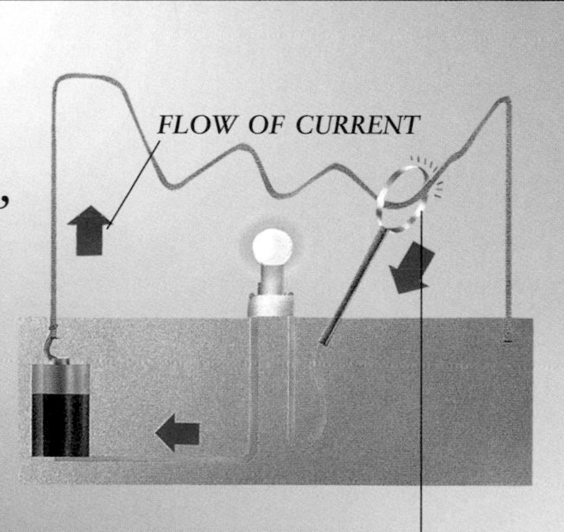

SWITCHING ON

WHAT YOU NEED
Styrofoam boards
Modeling clay
Two 6-V bulbs and sockets
Four 1.5-V batteries
Insulated wire
Thumbtacks
Paper clips

A SWITCHED CIRCUIT is a circuit in which the flow of electricity is controlled by a switch. When the switch is open, or off, there is a gap in the circuit and the electricity doesn't flow. Electricity flows when the switch is closed, or on. In this project you can use switches to send messages to your friends using flashing lights to make the letters in code.

SECRET MESSAGES

1 *Put a lightbulb in each socket and attach the sockets to separate styrofoam boards. Connect the two sockets to each other with a long wire.*

2 *On each board, connect the sockets to a thumbtack holding a paper clip. Add another thumbtack within reach of each paper clip. These will be your switches.*

3 *On each board, connect the second thumbtack to one end of a battery. Now connect the other ends of the batteries to each other with a long piece of wire. Make sure that the two ends of the batteries that you are connecting are different. Otherwise, electricity will not flow.*

MORSE CODE

You can send messages to a friend using Morse Code, by making the bulbs flash on and off quickly for dots and more slowly for dashes.

a	• –	*s*	• • •
b	– • • •	*t*	–
c	– • – •	*u*	• • –
d	– • •	*v*	• • • –
e	•	*w*	• – –
f	• • – •	*x*	– • • –
g	– – •	*y*	– • – –
h	• • • •	*z*	– – • •
i	• •	*1*	• – – – –
j	• – – –	*2*	• • – – –
k	– • –	*3*	• • • – –
l	• – • •	*4*	• • • • –
m	– –	*5*	• • • • •
n	– •	*6*	– • • • •
o	– – –	*7*	– – • • •
p	• – – •	*8*	– – – • •
q	– – • –	*9*	– – – – •
r	• – •	*0*	– – – – –

WHY IT WORKS

When the paper clips are in contact with the thumbtacks, the circuit is complete and electrons can flow, lighting the bulbs. To send a signal down the wire, the sender must raise and lower one of the paper clips to open and close the circuit. This is shown by the lightbulbs going on and off.

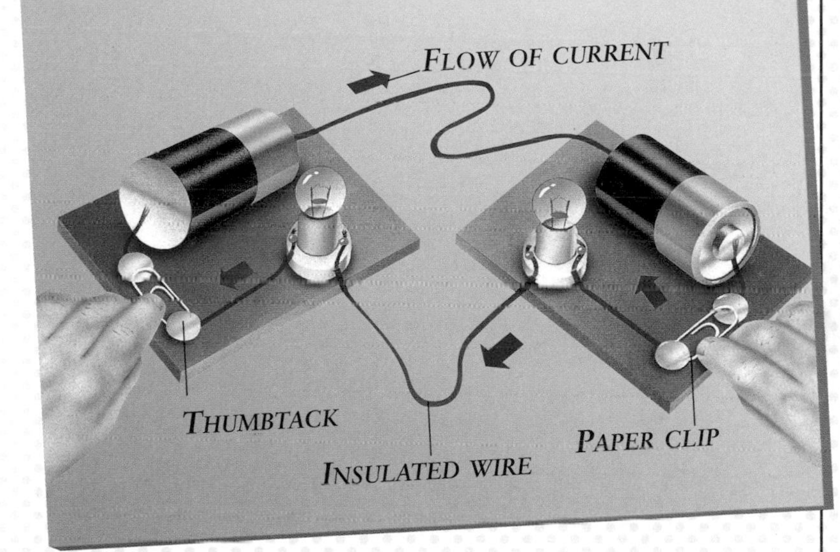

4 When the paper-clip switches are touching the thumbtacks, the bulbs will light up. Keep one switch closed, and open and close the other to send a signal down the wire.

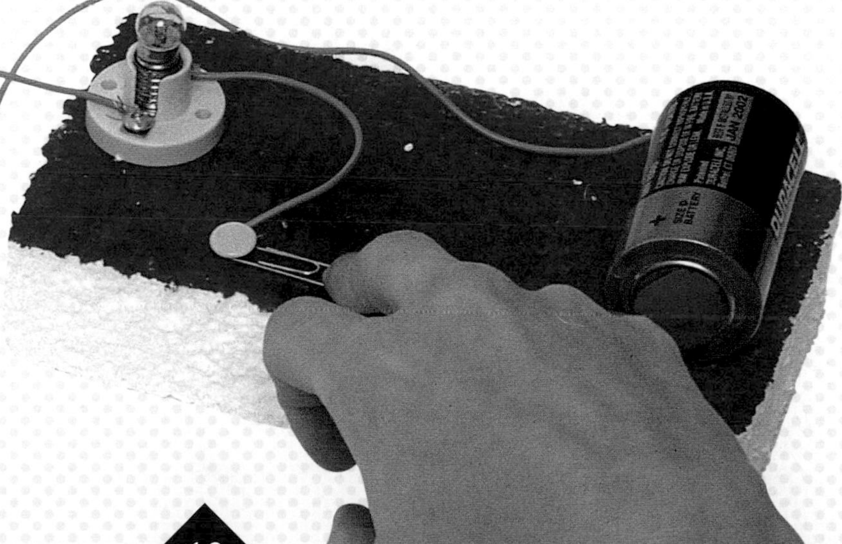

*T*WO-WAY SWITCHES

WHAT YOU NEED
Thumbtacks
Thick board
Two plastic lids
Paper clips
Two 1.5-V batteries
6-V lightbulb in socket
Insulated wire
Modeling clay

TWO-WAY SWITCHES ALLOW YOU TO TURN AN appliance, such as a light in your house, off and on from two different places. You may have a stairway light that you can turn on and off at the top and at the bottom of the stairs. Build your own two-way switch in this experiment and see how it works.

1 Make a hole in the top of a lid. Push the wires that you have connected to a bulb through a hole.

2 Open a paper clip and tape one end inside the lid so the other end sticks out. Attach one wire to the paper clip and the other to the batteries (below).

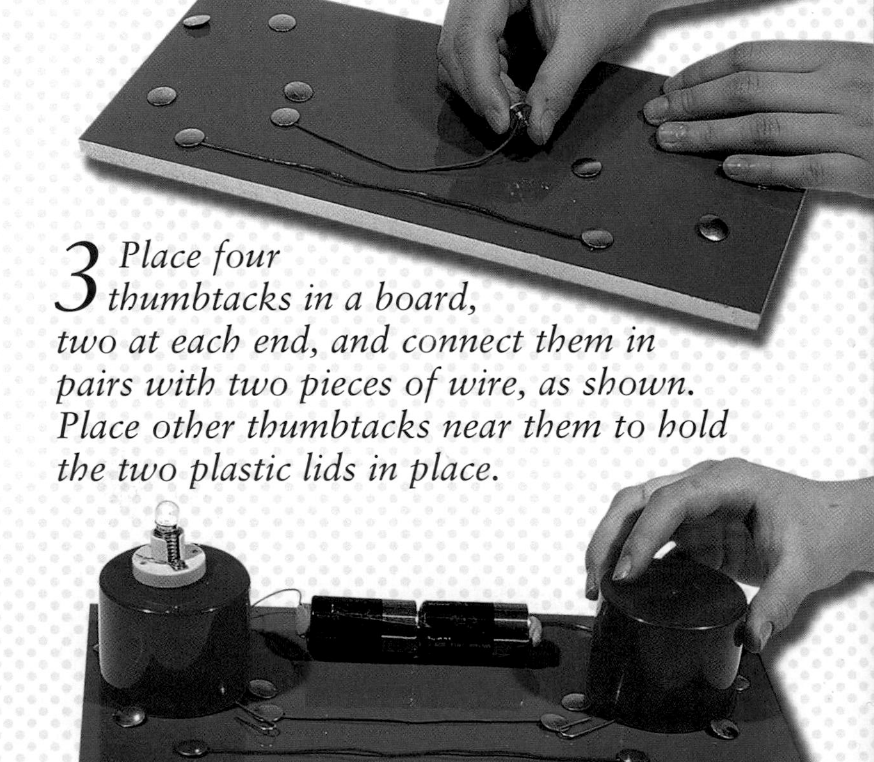

3 Place four thumbtacks in a board, two at each end, and connect them in pairs with two pieces of wire, as shown. Place other thumbtacks near them to hold the two plastic lids in place.

4 Form a switch with a paper clip in the other lid in the same way. Connect it to the other end of the batteries.

LOWERING SWITCHES

There are other uses for two-way switches. A person in a wheelchair may need a light switch lower down the wall at a height they can reach. Design a two-way switch circuit that would suit this purpose.

WHY IT WORKS

A two-way switch allows you to turn a light off and on in two different places. This is because there are two possible pathways for the electricity to flow along. These are the two wires (blue and red) you built into your circuit. For electricity to flow, both switches must be turned to the same pathway, either the blue or the red wire.

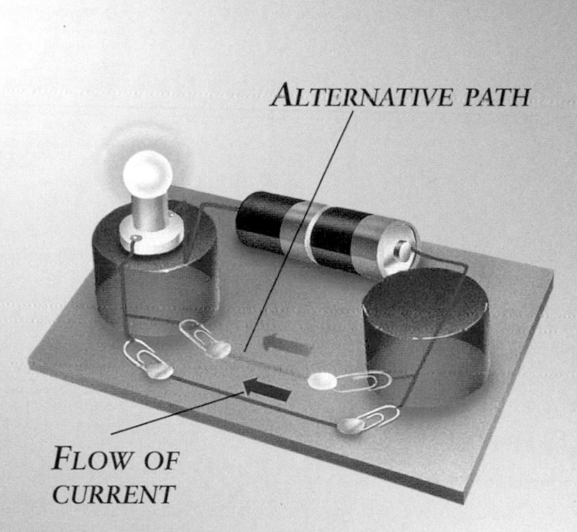

ON AND OFF

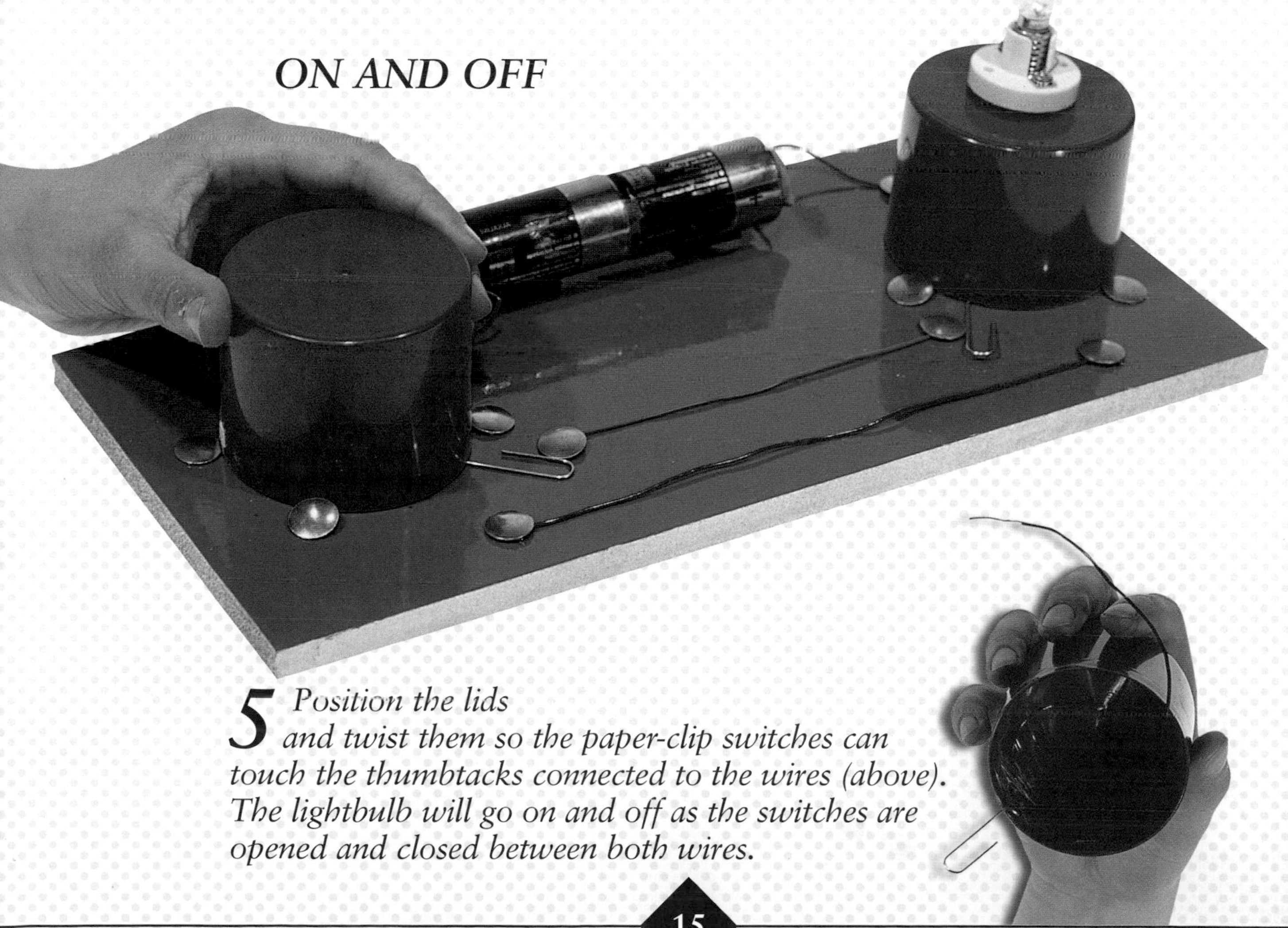

5 Position the lids and twist them so the paper-clip switches can touch the thumbtacks connected to the wires (above). The lightbulb will go on and off as the switches are opened and closed between both wires.

LIGHTING THE DARK

WHAT YOU NEED
Glue
Lightbulb in socket
Insulated wire
Colored cardboard
Brass fasteners
Battery

IN 1879, AMERICAN INVENTOR Thomas Edison made the first electric lightbulb. For the filament, the part that glows when electricity is passed through it, he used a piece of cotton thread heated to a black strip. He removed the air from the lightbulb and turned on the current. The lightbulb glowed. By 1913, a metal called tungsten was used as the filament.

ELECTRONIC QUIZ

1 Make up some questions and answers, and write each on a separate piece of paper.

2 Glue the questions in one column to one side of a piece of cardboard. Glue the answers in random order in another column on the cardboard.

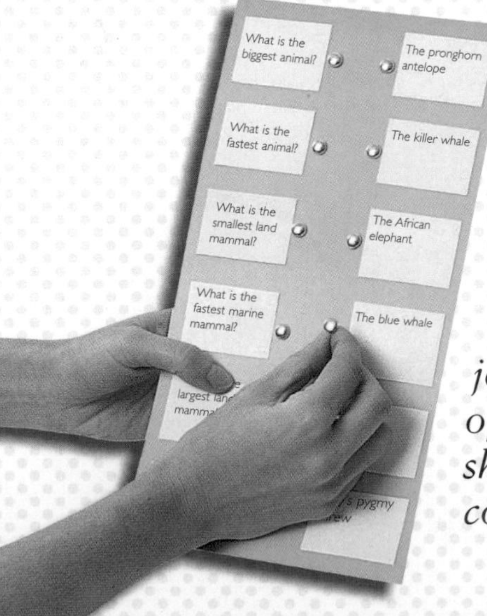

3 Push a brass paper fastener through the cardboard next to each question and each answer.

4 On the back of the cardboard, join the brass fastener of each question with a short length of wire to its correct answer.

5 Make a circuit with more wire, a battery, and a lightbulb. Leave the ends of the wires free.

TRAFFIC LIGHTS

Build a circuit with three colored lights to make traffic lights. Connect them together with switches so they can be turned on in different sequences depending on whether the traffic must stop or go.

WHY IT WORKS

Each question's paper fastener is connected by wire to the paper fastener of the correct answer. By touching the wires of your circuit to a question and its answer, you complete the circuit, so electricity flows and the bulb lights up. If the answer is wrong, the circuit is not completed, and the bulb will not light up.

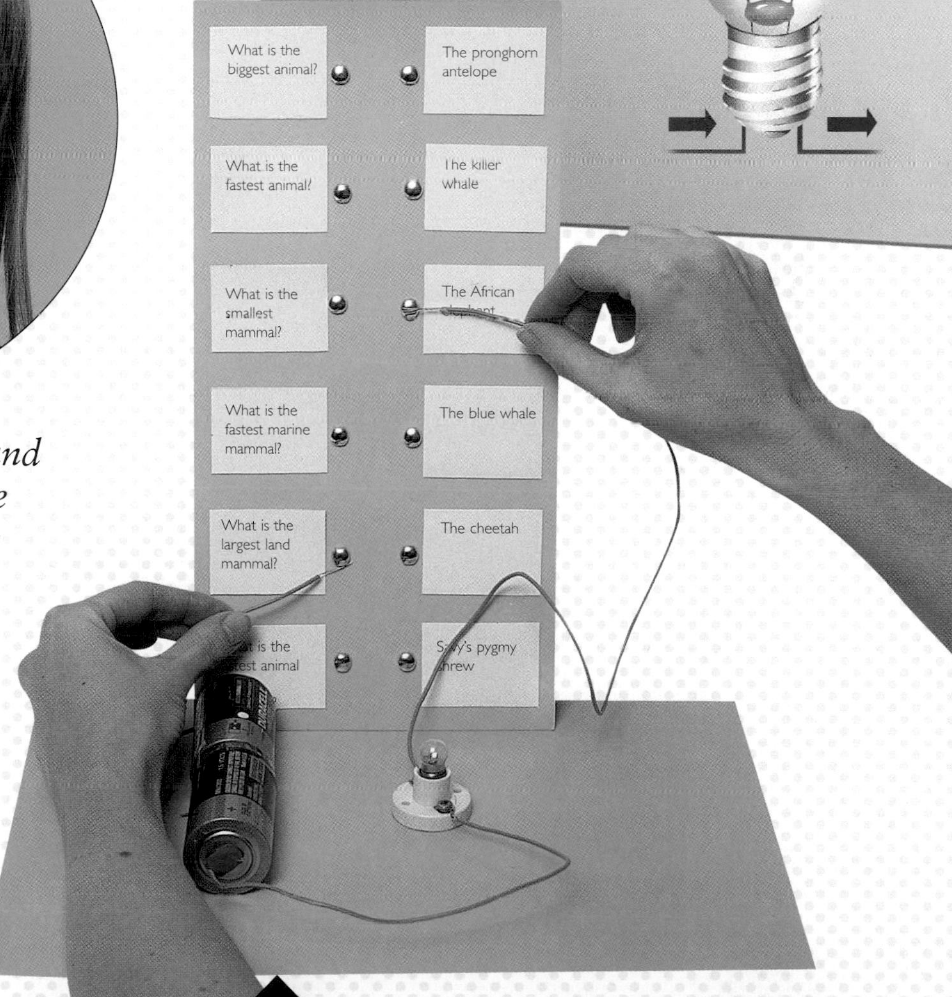

6 Ask a friend a question and let him or her choose one of the answers. With the free ends of the wires in your circuit, touch the paper fasteners next to the question and the answer your friend gives. If the bulb lights up, the answer is correct.

CONDUCTORS

SOME MATERIALS WILL NOT ALLOW ELECTRICITY to flow along them. They are called insulators. Other materials will let a current pass through them. They are called conductors. We need conductors to make electric wires and circuits, while insulators are important in protecting us from dangerous electric currents. This project will show you the difference between conductors and insulators.

WHAT YOU NEED
Thick board
Aluminum foil
Insulated wire
Adhesive vinyl
Two batteries
Lightbulb in socket
Nail

3 Before you stick down a final conducting path, make a hole near the edge of the board and insert the end of some insulated wire.

ELECTRIC MAZE

1 Cut a piece of aluminum foil the same size as your board. Cover the foil with a sheet of clear adhesive vinyl.

2 Design your maze on the board, and cut out strips of the vinyl-covered foil to fit your paths. Attach them plastic side up. These are your insulated paths.

4 Attach the other end of the wire to a battery. To the other terminal of the battery, attach a wire leading to a lightbulb in a socket. Attach another piece of wire to the other side of the socket and put a nail on the end of it.

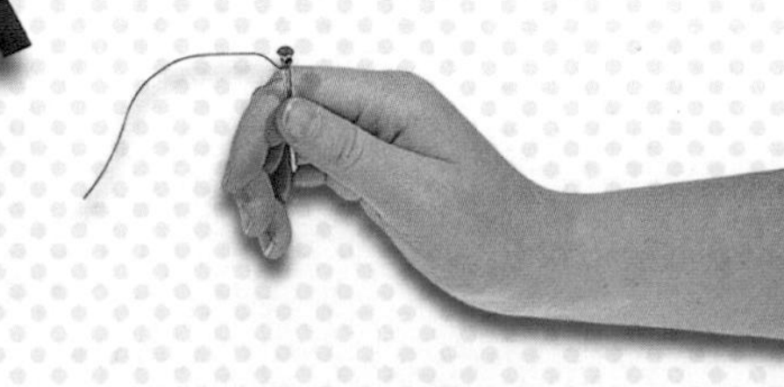

TESTING

You can test other materials to see if they are conductors using your circuit. Touch the free ends of the wires to the ends of objects made from different materials, such as erasers and spoons.

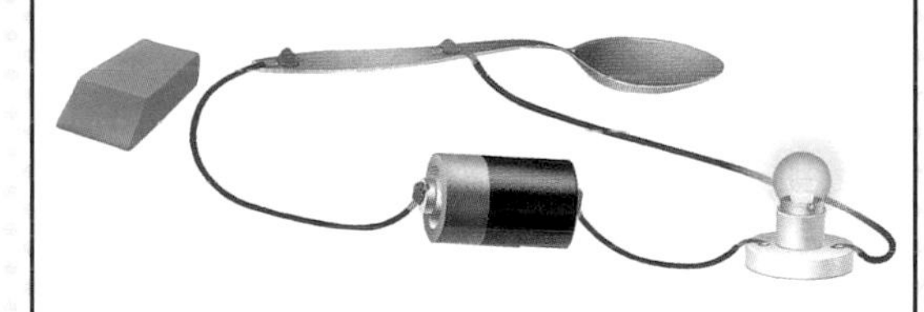

WHY IT WORKS

When the nail touches the plastic, the lightbulb goes out. The vinyl is an insulator and blocks the current. The electrons in an insulator are not free to move as they are in a conductor, such as aluminum foil, so the current doesn't flow.

5 *Cut out your final conducting path and stick it to your board, foil side up. Make sure the wire touches the foil at one end. Let your friends find their way through the maze.*

POOR CONDUCTORS AND RESISTORS

WHAT YOU NEED
Lightbulb in socket
Insulated wire
Thick board
Cardboard
Paper clip
Pencil lead
1.5-V batteries
Thumbtacks
Aluminum foil

NOT ALL CONDUCTORS ARE EQUAL. An electrical current can pass through some more easily than others. The thinner the wire, the higher the resistance, like the slower flow of water through pipes of smaller widths. Resistance also varies with length. In the experiment below you can test different conductors to see if they have high or low resistance.

RESISTANCE

1 Set up a circuit like the one shown, using insulated wire, thumbtacks, and a lightbulb in a socket. Get some lead from a mechanical pencil, and cut out two pieces of cardboard to rest it on.

2 To reflect the light,make a shade from a cardboard disk and aluminum foil.

3 Cut a slit in the disk and glue it into a cone shape. Cut a hole in the center of the cone, then place it over the lightbulb.

WOOD

Replace the lead in the experiment with a wooden skewer that has been soaked in salt water overnight. The salt in the water should allow the wood to conduct electricity. As the wood begins to dry out, the resistance will increase until eventually the wood will stop conducting.

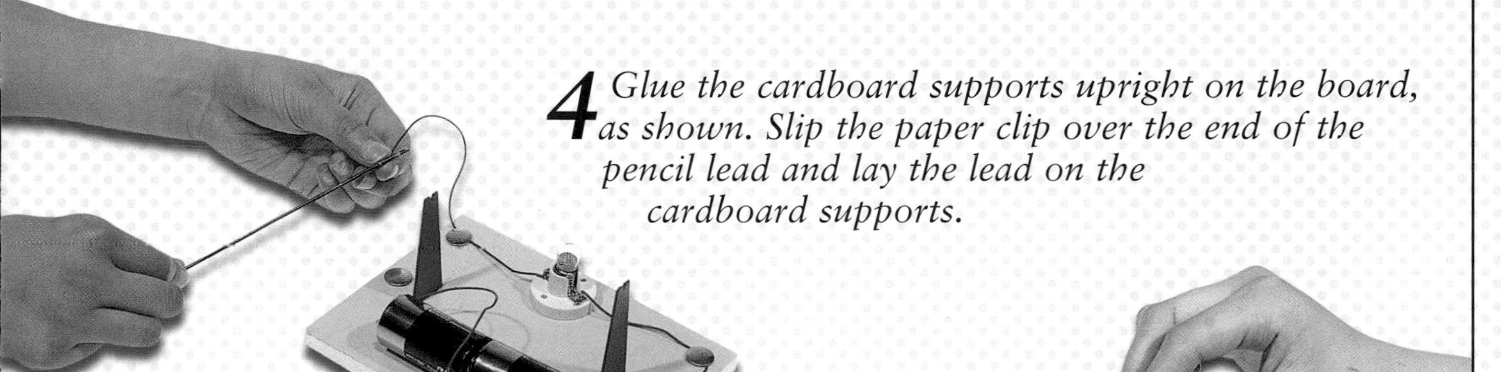

4 Glue the cardboard supports upright on the board, as shown. Slip the paper clip over the end of the pencil lead and lay the lead on the cardboard supports.

5 Attach the end of one of your wires to the end of a battery. Connect another piece of wire to the other end of the battery. Add the paper clip over the lead to the end of this wire. Attach the end of your other free wire to the end of the pencil lead.

WHY IT WORKS

Resistance increases the farther the current has to travel. As the paper clip moves toward the battery end, the distance the current has to move is shortened, so there is less resistance and the light bulb glows more brightly.

VOLTAGE AND CIRCUITS

IF YOU PUT MANY LIGHTBULBS in a circuit in a line, one after the other, they are said to be in series. If one of the lightbulbs goes out, they all go out. Streetlights in the early 1900s were set up like this, and streets were plunged into darkness when one of the lights failed. The solution was to put them in parallel, so the current didn't have to go through one lightbulb to get to another. Streetlights today are in parallel circuits, so that if one lightbulb fails, the others will continue to glow.

WHAT YOU NEED
Two large boards
Thumbtacks
Insulated wire
3-V lightbulbs in sockets
1.5-V batteries

WHY IT WORKS

A series circuit uses one path to connect the lightbulb to the battery. If two batteries are used, the lightbulb glows twice as brightly. Two lightbulbs in a series circuit glow less brightly than one. A parallel circuit has more than one path for the current. Each lightbulb receives the current at the same force, or voltage, no matter how many lightbulbs there are in parallel. If a lightbulb burns out, the others continue to glow because their circuits are not broken.

SERIES AND PARALLEL

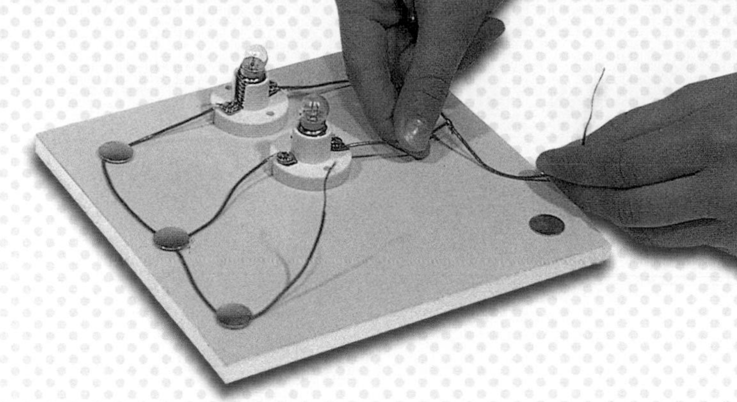

1 Set up a circuit with two lightbulbs in parallel, connecting the bare ends of the wires to thumbtacks as shown. Attach the free ends of the wire to the ends of a battery. Both lightbulbs glow with equal brightness.

2 Now remove one of the lightbulbs. The other should still be lit.

3 Now set up a circuit with two batteries connected end to end, in series. Connect them to a single lightbulb. How brightly does it glow? Take one of the batteries out. How brightly is the lightbulb glowing now?

OTHER CIRCUITS

Using the parallel circuit that you made in the project, replace the bulb in the middle of the circuit with a battery, as shown here. How does this affect the lightbulb? Does it glow brighter or dimmer? Or does it not glow at all?

ELECTRICITY AND IONS

AN ELECTRICAL CURRENT can pass through liquids, such as a salt solution, causing a chemical reaction. This is called electrolysis. Two metal plates, called electrodes, deliver the current to the solution (electrolyte). Electrolysis is used to coat metallic objects with a thin layer of a more expensive, attractive, or hard-wearing metal. This process is called electroplating.

WHAT YOU NEED

Glass jar of salt water
Copper coin
Paper clip
Two batteries
Insulated wire
Modeling clay

ELECTROPLATING

1 Connect the two batteries with the unlike terminals touching. Connect insulated wire to the free terminals. Attach the copper coin to the wire from the positive battery terminal.

2 Attach a paper clip to the wire from the negative battery terminal. Fill the jar with water and place the coin and the paper clip in the water. They will act as the electrodes.

SALT AND VINEGAR

Try the experiment again using a solution of salt dissolved in vinegar. Do you notice any difference? Does anything happen to the paper clip? Add more batteries in parallel to increase the pressure of the current.

3 Watch closely to see what happens. Are there bubbles forming? Leave the coin and the paper clip for a few minutes before taking them out. Are there any color changes? Put them back in the water for a while longer. Can you see any more changes?

WHY IT WORKS

The electricity flows through the solution as charged particles called ions. Copper ions carry the positive charge toward the negative paper clip, where they pick up electrons and are deposited on the paper clip as a thin layer of copper.

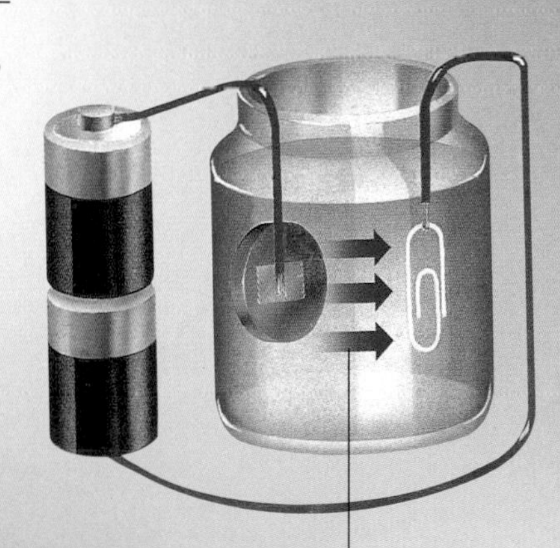

CURRENT FLOWS THROUGH THE SOLUTION, CARRYING COPPER IONS TO THE PAPER CLIP

MAGNETISM

IN THE 1820S A DANISH SCIENTIST NAMED HANS CHRISTIAN OERSTED found a link between electricity and magnetism. He noticed that a magnetic compass needle was deflected when an electric wire was held near it. He soon realized that an electric current flowing through a wire creates a magnetic field. You can make an electromagnet in the project below.

WHAT YOU NEED
Styrofoam boards
Nail
Insulated wire
Thumbtacks
Two 1.5-V batteries
Modeling clay
Paper clip
Table-tennis ball

ELECTROMAGNETIC FACE

1 *Push a nail through the center of a styrofoam board. Wrap wire around the nail at least 20 times, leaving the two ends the same length.*

2 *Cut two triangular pieces of styrofoam to support the board in a sloping position.*

3 *Attach the triangles as shown. Pierce a small hole in the side of one of the triangles so a paper clip will fit through.*

WHY IT WORKS

The current flows through the wire coiled around the nail and turns it into a magnet. The clown's nose is held in place because the thumbtack is attracted to the magnetic field. When the electricity is turned off, the magnetic field disappears and the nose falls off.

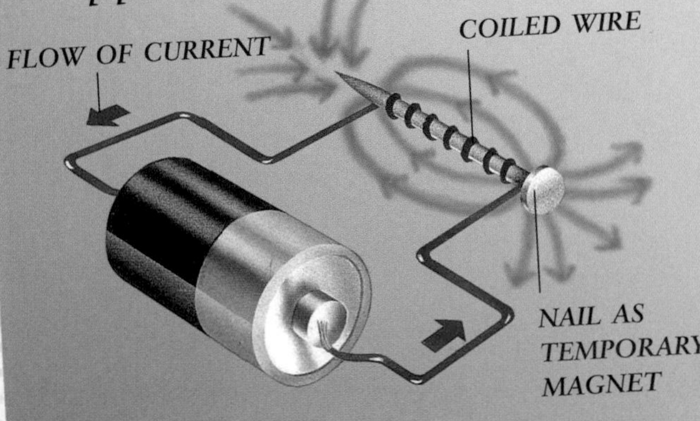

4 Draw a clown's face on a second piece of styrofoam board. Do not draw a nose on the face. Color the face and cut it out to fit over your sloping board.

5 Color a table-tennis ball red to make a clown's nose and push a thumbtack into it.

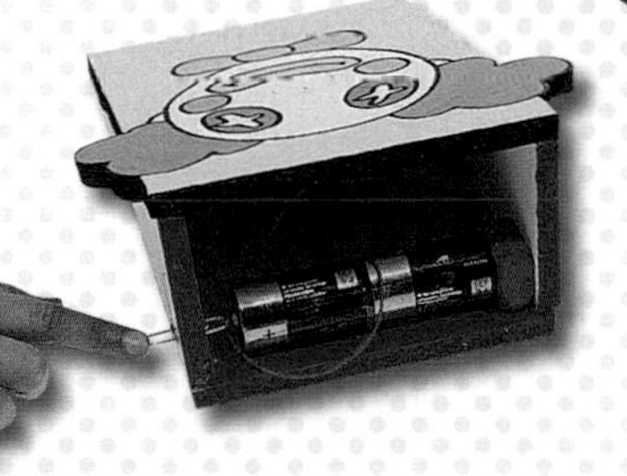

6 Position two batteries inside the shape as shown. Connect one of the wires from the nail to one end of the batteries. Connect the other wire to the paper clip. Glue the face board to the top of the shape.

7 Push the paper clip through the hole so that it touches one terminal of the batteries. Position the red nose on the clown's face.

8 While the paper clip is touching the batteries, the circuit is complete and the nose will stick to the clown's face.

MAGNETIC LINES

Pierce the center of a piece of paper with a piece of wire. Sprinkle some iron filings on the paper. Attach both ends of the wire to two connected batteries. The iron filings will line up in concentric circles, showing the "lines of force" of the magnetic field.

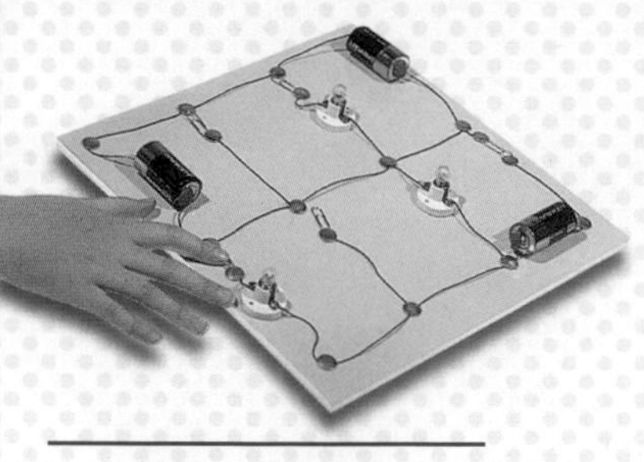

THE HOME CIRCUIT

HOMES TODAY CONTAIN PARALLEL CIRCUITS. One of these circuits has sockets in the walls. Lights can be run off a separate circuit. All lights and appliances are connected in parallel so that everything operates at the same voltage; turning things on and off does not change the voltage to other appliances. Experiment with more complex circuits in the project below.

WHAT YOU NEED

Large board
Three bulbs
Three 1.5-V batteries
Insulated wire
Thumbtacks
Modeling clay
Paper clips

SWITCHING BATTERIES

Experiment with the way the batteries are facing. Turn one around and see how it affects the rest of the circuit. Can all the lightbulbs be lit up now? Do the lightbulbs glow as brightly as before? Remember that electrons flow from negative to positive only.

SWITCHES AND LIGHTS

1 Place batteries at three corners of the board. Make sure that unlike terminals are facing each other. Attach wires using modeling clay.

2 Connect the lightbulbs to the batteries using short lengths of wire running between thumbtacks as shown. Leave gaps in the circuits for switches made from paper clips and thumbtacks.

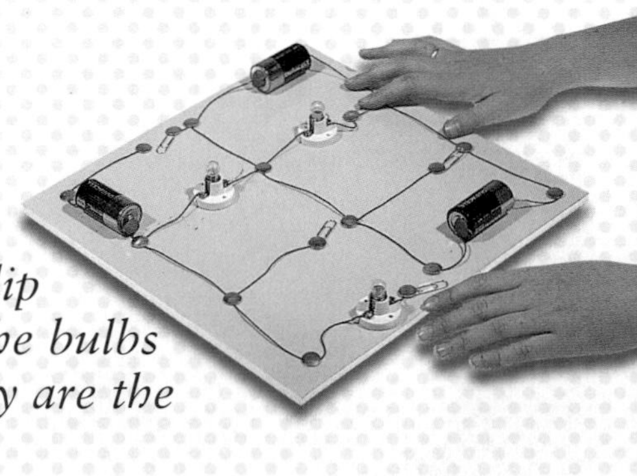

3 Close the paper-clip switches and see the bulbs light up. How brightly are the lightbulbs glowing?

4 By opening and closing certain switches, can you have just one lightbulb on? Experiment with the switches to see if you can get only two to light up. Do the lightbulbs glow at different degrees of brightness?

WHY IT WORKS

The flow of electrons is controlled by closing and opening the switches. A lightbulb stops glowing when the electrons no longer flow through it. All the lightbulbs will glow when every switch is connected (closed).

GLOSSARY

CONDUCTOR (kun-DUK-ter) A material that will allow electrons to flow through it. Examples include metals and carbon. *Experiment with different conductors on pages 18-19.*

ELECTRIC CIRCUIT (ih-LEK-trik SUR-kut) The path around which an electric current flows. It includes a source of electricity, such as a battery. *You can make a simple circuit in the project on pages 10-11.*

ELECTRIC CURRENT (ih-LEK-trik KUR-ent) The movement of electrons along a wire. *Make a current flow in the project on pages 10-11.*

ELECTRODE (ih-LEK-trohd) When sending an electric current through a liquid, such as a salt solution, the electricity is carried into the liquid by two conductors called electrodes. *See how electrodes are used on pages 28-29.*

ELECTROMAGNET (ih-lek-troh-MAG-net) An iron or steel object surrounded by a coil of wire that acts like a magnet when a current flows through the wire. *Make your own electromagnet in the project on pages 26-27.*

GLOSSARY

ELECTRON (ih-LEK-tron) One of the tiny particles that make up atoms. Electrons have a negative charge. Whether moving or static, electrons are what we call electricity. *Find out more about electrons in the project on pages 8-9.*

ION (EYE-un) A charged particle in a fluid. *Use ions to carry electrical charge through a salt solution in the project on pages 28-29.*

INSULATORS (IN-suh-layt-urz) Matter that does not conduct heat or electricity. *Experiment with insulators on pages 18-19.*

STATIC ELECTRICITY (STA-tik ih-lek-TRIH-suh-tee) A type of electricity that forms on the surface of certain materials that are rubbed together. *Experiment with static electricity in the projects on pages 6-9.*

VOLTAGE (VOHL-tij) The force that pushes electricity through a wire, similar to the pressure of water in a pipe. *Experiment with voltage on pages 22-23.*

INDEX